# Windmills in England

*Pakenham Mill, Suffolk*

# *Windmills in England*

a study of their origin, development and future

by Rex Wailes

*London: Charles Skilton Ltd*

*First published 1948, by The Architectural Press*
*Reprinted 1975 by Charles Skilton Ltd*
*90 The Broadway London SW19*

TO THE MEMORY OF

THOMAS HENNELL, R.W.S.

with whom I had hoped to write this book

Printed in Great Britain by
Lowe & Brydone (Printers) Ltd, Thetford, Norfolk

# Preface

*We said withal there was a terrible flash of anger in Samson; witness his address to old Herbert the Dean, who in a too thrifty manner has erected a windmill for himself on his glebe lands at Haberdon. On the morrow, after mass, our Lord Abbot orders the Cellerarius to send off his carpenters to demolish the said structure* brevi manu, *and lay up the wood in safe keeping. Old Dean Herbert, hearing what was toward, comes tottering along hither, to plead humbly for himself and his mill. The Abbot answers: 'I am obliged to thee as if thou hadst cut off both my feet! By God's face,* per os Dei, *I will not eat bread till that fabric be torn in pieces. Thou art an old man and shouldst have known that neither the King nor his Justiciary dare change aught within the Liberties without consent of Abbot and Convent; and thou hast presumed on such a thing? I tell thee it will* not *be without damage to my mills; for the Townsfolk will go to thy mill and grind their corn* (bladum suum) *at their own good pleasure; nor can I hinder them, since they are free men. I will allow no new mills on such principle. Away, away; before thou gettest home again thou shalt see what thy mill has grown to!'—The Old Dean totters home in all haste; tears the mill in pieces by his own* carpentarii, *to save at least the timber; and Abbot Samson's workmen, coming up, find the ground already clear of it.*

*Easy to bully down poor old rural Deans, and blow their windmills away; but who is the man that dare abide King Richard's anger; cross the Lion in his path, and take him by the whiskers! Abbot Samson too; he is that man, with justice on his side . . .*[1]

*By the law of milling soke no mill was allowed to be set up in opposition to the manorial mill, because a mill was a complicated piece of machinery and, if too many of them had been erected, the survival of any one—in this case Abbot Samson's—might have been very doubtful. In the later Middle Ages, with the increase of population, the law became obsolete.*[2]

When in 1923 the late K. J. Tarrant asked me to photograph some of the Lincolnshire windmills for him, I little knew what it would lead to. He was a retired engineer who had been interested in windmills since, as a boy, he had been allowed to work the old postmill at Blackmore End, Essex, in the miller's absence. I was an engineering apprentice in Lincoln, and had a motor cycle, a quarter-inch ordnance map and a V.P.K. and used all three at week-ends. One day, while I was photographing the big six-sailed tower mill at Coleby Heath, now demolished on account of the nearby aerodrome, the miller asked me if I would like to have a look inside. He showed me over from bottom to top and we finally ended up just under the fantail, looking out over a stretch of country that is said to have been settled by the ex-service men of Roman times. Here in the mechanism of the mill was part of the history of mechanical engineering in the solid, just the thing to whet the appetite of a budding engineer with a taste for the historical side of his profession. Since that day I have tried to have a look inside every windmill I have come across, both at home and abroad.

In London in 1925 I joined the Newcomen Society for the Study of the History of Engineering and Technology. Here I met four men who were also keenly interested in windmills. They were the late Mr. E. Lancaster Burne, a consulting engineer in London, an expert on the design of windmill sails and himself a patentee of a design that anticipated some of the improvements now to be found on windmill sails in Holland, the late Mr. H. O. Clark of Norwich,

[1] Thomas Carlyle (*Past and Present*, Book II., Chapter 15, 1843). Translated from Jocelyn de Brakelond's 'Chronicon', Chap. 43.

[2] *History of Corn Milling*, Vol. II., 1898. Those who are interested in the history of corn milling in general and of windmills in particular cannot do better than consult Bennett and Elton. Their book, much quoted, seldom acknowledged and never surpassed, is not so satisfactory from the technical point of view.

whose knowledge of the windmills of Norfolk was exhaustive and whose superb model of Sprowston post mill is now in the Science Museum, London. Mr. John Russell, director of Union Mills, Cranbrook, who is in addition a practical engineer and millwright, and Mr. A. Titley, a consulting engineer in Birmingham, a founder and first president of the Newcomen Society, whose paper to the Society on windmills is a classic. These four helped and encouraged me to collect and collate data and to make notes and photographic records of the interiors of windmills as well as their exteriors.

In 1929 the Society for the Protection of Ancient Buildings became interested in windmills and in 1931 formed a Windmill Section, which helped with the repair and preservation of a large number of windmills. I became its honorary technical adviser, and came into contact with millwrights some of whom are referred to in these pages, and to them, too, I owe a debt of gratitude for the help and advice that they have unstintingly given.

The reader will by now have gathered that it is the technical side of windmills that has interested me most, and has lead to a good deal of work in the field, taken me into many unusual, interesting and beautiful places that I should not otherwise have visited, and brought me into contact with many kind, delightful and hospitable folk. But the technical is not the only angle from which windmills can be studied. Thomas Hennell's drawings have shown that it is possible to make a fine picture of a windmill and yet be accurate enough to satisfy millwrights and millers. Architects have interested themselves in the construction of windmills, photographers have found them particularly pictorial, while local historians have, in windmills and millers of the past, subjects of more than ordinary interest. It is not generally realised that, before the days of steam power, almost every parish had its own wind or water mill, or both and often more than one; that the miller was as important a member of the community as the blacksmith, and that the history of the country mill from feudal times to the present day presents a facet of the history of the countryside as a whole which might well be more fully explored by social historians than it has been so far. So the very brief survey of windmills contained in this book is but an introduction to one aspect of a subject which ought to attract people with a wide range of tastes; and if it can stimulate further interest in windmills in any direction it will have fulfilled the object for which it was written.

REX WAILES.

*London,* 1948.

## Note to this Reprint

The dates and other references must be taken as relating to the 1948 edition.

## Acknowledgments

*The major contents of this book were originally published as a special number of* The Architectural Review *in September 1945, and thanks are due to the Editor for permission to include this material. Photographs of figures 2, 3, 5, 11, 18, 49, 50, 52, 55, 58, 67, 76, 77, 82, 86, 88, 90, were taken for the author by Mr. Hallam Ashley, F.R.P.S. who also took figures 65 and 66. Major T. B. Paisley has kindly permitted the reproduction of 16 and 71; while 79 is from a photograph by the late Mr. K. J. Tarrant in the possession of the author. The design on the title page is by Mr. Vincent Lines and the four drawings, figures 7 to 10 are by Mr. R. O. Fyson.*

# Contents

# Illustrations

2

*'Along the flat horizon there arise the frequent venerable towers of churches. He sees at the end of airy vistas the revolution of the windmill sails. He may go where he pleases in the future, he may see alps and pyramids, and lions, but it will be hard to beat the pleasure of that moment. There are indeed few merrier spectacles than that of many windmills bickering together in a fresh breeze over a woody country, their halting alacrity of movement, their pleasant business making bread all day with uncouth gesticulations, their air gigantically human, as of a creature half alive, put a spirit of romance into the tame landscape. When (the traveller) sees them first he falls immediately in love, and from that time forward windmills keep turning in his dreams.'*

R. L. Stevenson, *'The Foreigner at Home'*.

**preamble**

"Fitness or the proper adaptation of means to an end is the great source of the relative beauty of forms." Thus, Sir Thomas Dick Lauder in 1842, quoted in *The Architectural Review* in April, 1944. It explains the attraction of windmills to many laymen. But why, if so many like them, do so few know about them? Their origin, development and future in this country should be the concern of a much wider circle than is at present the case. To the average town-dweller the windmills are merely part of the landscapc and mean nothing in particular unless his attention has been especially directed to them. Yet the appeal of the windmill is many sided—æsthetic, economic, mechanical, structural and historical—and like the steam locomotive "there is no mistaking its purpose or the means by which it does its work".[1]

**distribution and decline**

In their hey-day there may have been as many as ten thousand windmills at work in England at one time. To-day there are less than fifty including the marsh mills used for drainage on the Broads and in the Fens.

The location of this once mighty army was conditioned by two main considerations: first by the existence of good corn country and secondly the necessity to drain land. The corn windmills were naturally most plentiful where land was fairly level and water power lacking; that is, broadly speaking, east of a line drawn from Newcastle to Portsmouth. A fair number were to be found scattered over the remaining Midlands and Southern Counties, and they were fairly numerous also in Somerset, Wirral, the Lancashire Fylde and Anglesey, the granary of Wales. To-day they are to be found at work only in Lincolnshire, Nottinghamshire, Norfolk, Suffolk, Cambridgeshire and Kent, while Yorkshire, Essex, Derbyshire, Leicestershire, Surrey, and Sussex have only one each. Of the two thousand or more that once drained the Fenland less than half a dozen dwarf examples remain; the rest have been ousted by steam, oil and electricity. The corn windmill has been driven out of existence by other sources of power at the very time when it was at its best. In addition, agricultural depression, the depopulation of the countryside, restrictive legislation and the competition of the big port mills

[1] W. J. Gordon: *Our Home Railways.*

A

have killed the trade of the windmill to such an extent that between the two German Wars they went out of use at the rate of about one a month. As to the millwrights capable of repairs and renewals to windmills there are not a dozen firms left in the country.

**origin**

The windmill was unrecorded in Western Europe at the time of Domesday Book, and the mills mentioned therein were either watermills or cattle mills. The earliest references to windmills in England occur late in the twelfth century, when they were evidently a curiosity. Each is specifically distinguished as 'molendinum ad ventum' as opposed to the usual 'molendinum'. Under the laws of milling soke mills could only be erected by the lord of the manor, and all corn had to be ground in the manorial mill. This encouraged the erection of costly water and windmills by eliminating wasteful competition and price cutting. It was responsible for an early and amusing account of the infringement and of the subsequent demolition of the offending windmill in *The Chronicles of Jocelyn*, quoted in the preface.

One hundred years later the windmill begins to appear in illuminated manuscripts; a distinction only given to familiar objects.[1] Mediæval representations of windmills occur also in other works of art. Thus there are a carved bench-end at Thornham, Norfolk, a memorial brass at King's Lynn, Norfolk, a stained glass quarry at Stoke-by-Clare, Suffolk, a boss in Norwich Cathedral, and a carving in granite at Launceston, Cornwall, to name but a few, 2, 3, 4, 5, 6.

**the post mill**

The windmill as we know it seems to have appeared in Western Europe at the beginning of the twelfth century; but early references are scarce, many are apocryphal and they do nothing to throw light on the origin of the windmill. The early representations make it clear, however, that the earliest European windmill is of the type known as the 'post mill', 2, also page 7. A box-like timber body, carrying the sails and containing the machinery is mounted on a massive upright oaken 'post', 12, 13, on which it can turn. The weight is taken by diagonal 'quarter-bars' mortised into it about half-way up and held to the two horizontal 'cross-trees' below by bird-mouth joints. To steady the post it is quartered over the cross-trees, which rest on piers but are not fixed to them in any way.

Socketed on top of the post like the top stroke of a T is the 'crowntree', on which the body of the mill is built and with which it is free to turn round. On page 5 is shown a postmill in section revealing the form of the structure. Two horizontal 'side girts' rest on the end of the crowntree; these carry a corner post at each end. The sides of the mill are formed by uprights and braces, the roof is framed-up in the usual manner and two stout horizontal 'sheers' run fore and aft on either side of the post below the bottom floor to support the central upright in the 'breast' of the mill and to steady the body of the mill about the post by forming a bearing or carrying a 'collar', which turns round it and prevents the mill from swaying. The whole framing of the mill is covered with weatherboards, which were occasionally plastered inside and, in England, often painted white. The early mills were lighted by shuttered openings in the weatherboarding; later, glazed windows were used and in

[1] The earliest known illustration of a windmill is to be found in the so-called *Windmill Psalter* in the Morgan Collection.

3

4

5

## WINDMILLS IN MEDIAEVAL ART

*3, boss in Norwich Cathedral cloisters. 4, carving in granite on the exterior of St. Mary Magdalene Church, Launceston, Cornwall. 2, 3 and 4 represent a mediaeval joke. A miller is returning to his mill with a sack of corn for grinding laid across his horse's back. Seeing his horse is tired he says 'a merciful man is merciful to his beast'. He shoulders the sack himself—but mounts the horse! 5, bench-end in Thornham Church, Norfolk. 6, stained glass quarry in Stoke-by-Clare Church, Suffolk. The headpiece on page 1 is from the Flemish memorial brass to Adam de Walsoken, St. Margaret's Church, King's Lynn, Norfolk.*

6

some cases were of considerable size. Access to the mill body is obtained by a ladder running from a platform at the tail of the bottom floor to the ground. The inclined windshaft in the top of the mill carries the sails, and the mill is turned into the wind by means of a 'tail pole' fastened to the sheers and passing through the ladder. The substructure is protected from the weather by a 'roundhouse', which has no structural significance though it serves as a useful storage space. In a very few cases the roundhouse has an iron-faced 'curb' on the top, on which run rollers fixed to the bottom of the lower floor of the mill body, in order to prevent it from pitching, 87. Roundhouses are usually of brick, with boarded and felted roofs; but examples were to be found in wood, stone, and even clay lump, while roofs were tiled, slated, or occasionally thatched. The roundhouse is not attached by its rafters to the post of the mill; the weatherboarding of the mill body is therefore usually extended downwards to form a petticoat, 81, which overlaps the top of the roundhouse roof. In some cases, however, a small round petticoat is provided, 77, while in others, where there is no roof to the roundhouse, a built-out petticoat itself forms the roof, 41 and 87. The most satisfying æsthetically were some in West Suffolk of brick and flint with tiled roofs, looking as though they had grown out of the ground on which they stood, 78. In most cases the roundhouse was a later addition; but in some nineteenth century East Suffolk post mills they are coeval with the mill, 79, contain three floors and are as high as the body of the mill proper, 80.

Early post mills were small, the pent roof was straight and the weatherboarding, which covered the framework, was wide. The only example of this sort surviving in England stands at Bourn in Cambridgeshire, 96. From its appearance I decided that it was of early seventeenth century date. After

## THE CONSTRUCTION OF WINDMILLS

*7. This marsh mill was built at Soham Mere, Cambridgeshire, for draining the fenland in 1867 by the grandfather of Mr. T. B. Hunt of Soham, 63. It was built on the site of an earlier marsh mill for two brothers, one of whom attained the remarkable age of 107. It is in the old tradition of the large drainage mills, immortalized beautifully, if inaccurately by Cotman, and is the very last of its kind in the Fens, the few remaining survivors being but 20 feet or so in height. The drive is simple. The 'brake wheel' on the 'windshaft' engages with a large 'trundle' 'wallower', on top of the upright shaft, which has pegs instead of cogs. At the lower end of the shaft a spur wheel drives a countershaft carrying a bevel pinion which engages with the 'pit wheel' on the 'scoop wheel' shaft. The scoop wheel which is 28 ft. 6 in dia. × 12 in. wide, raised water up into the main drain by means of paddles or floats and the arrangement of gearing resembles the drive of a watermill in reverse. The mill has been condemned and is to be demolished.*

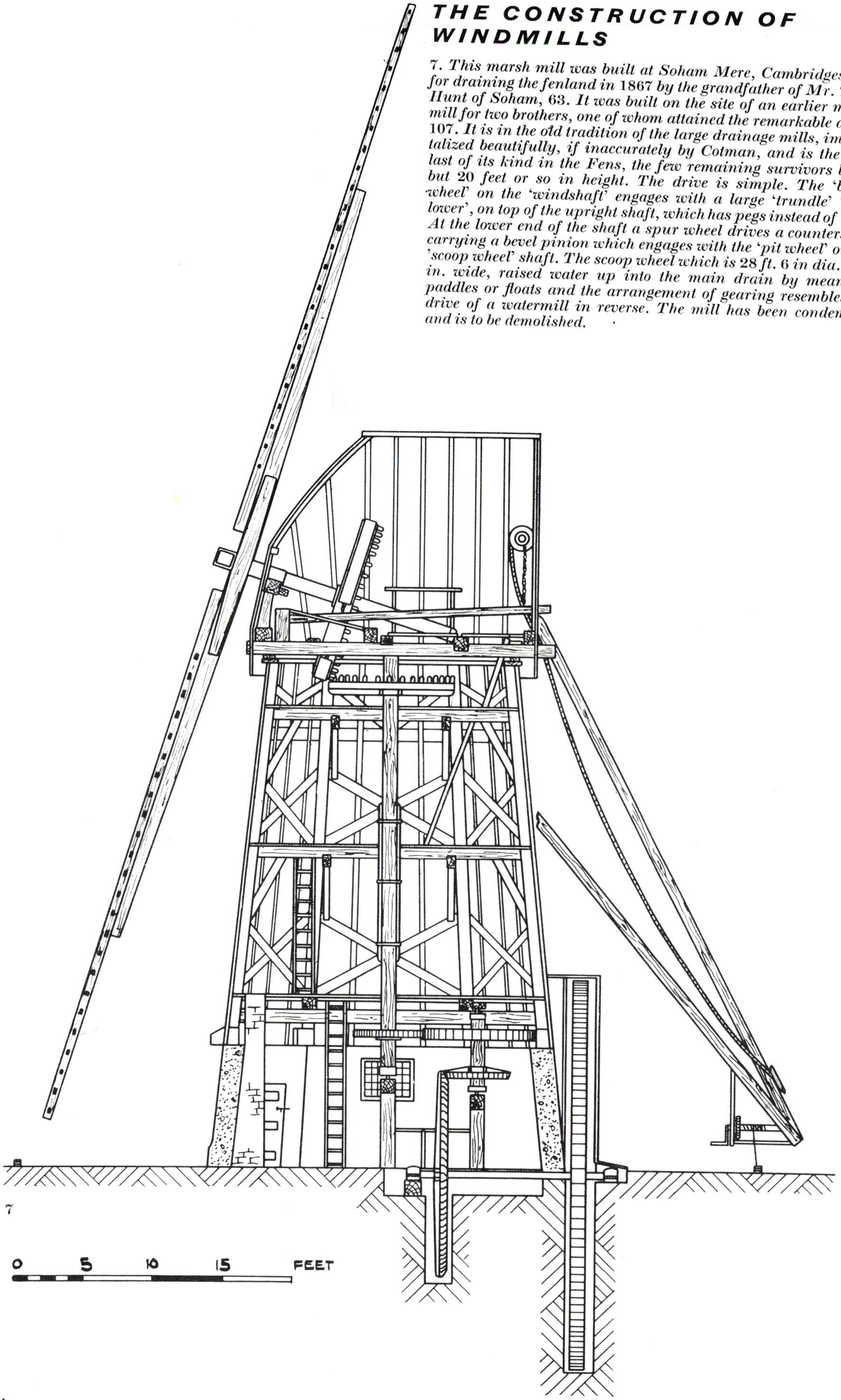


7

*Three more Cambridgeshire mills are here shown. The tower mill, 8, is at Swaffham Prior. It was built by Richard Fyson of Soham about the year 1858 on the site of an earlier fret mill and still stands, though not in use. The tower is of brick and flint, and plastered and tarred on the outside. The cap, like that of the smock mill, 9, which stood at North End, Burwell, is typical of Cambridgeshire practice. Internally the arrangement of the gearing in both mills is similar and the 'sackhoist', driven by a friction ring from below the wallower is plainly shown. The Burwell mill was known as the Busy Bee. The postmill, 10, which bears the date 1764, stands disused at Six Mile Bottom and was moved to its present position when the Cambridge to Newmarket railway line was built. It has two pairs of stones, front and rear, driven from the brakewheel and 'tailwheel' respectively. The sackhoist is driven by belt from a pulley in front of the brakewheel and the flour dressing machine from a spur wheel engaging with the inside face of the brakewheel cogs. The drawings on pages 4 and 5 are to the same scale.*

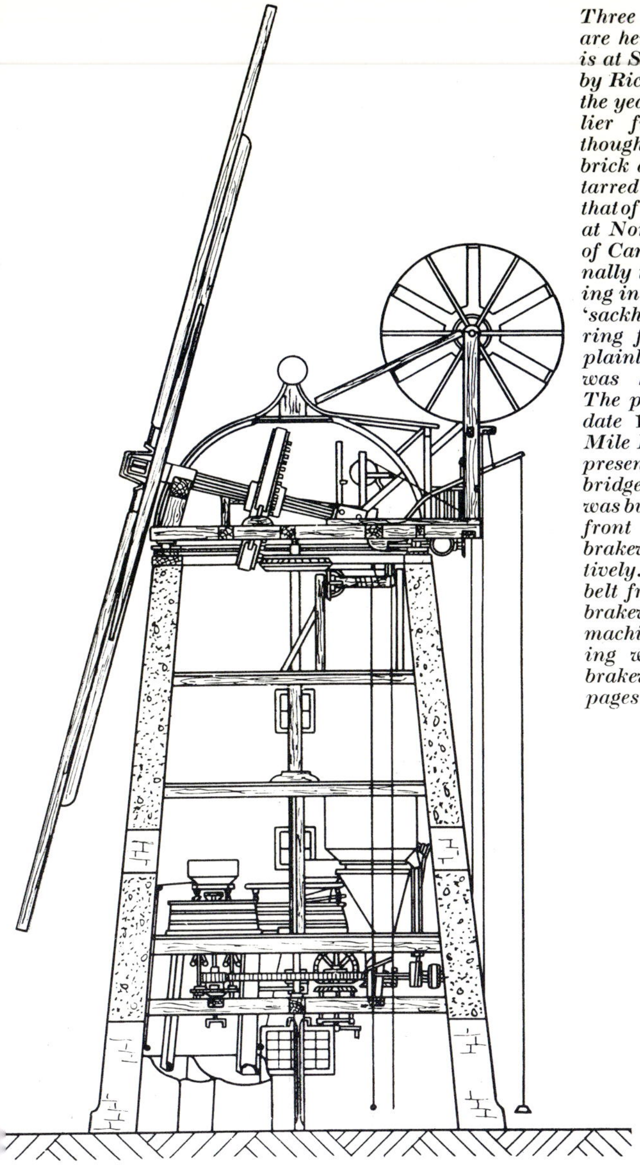

8

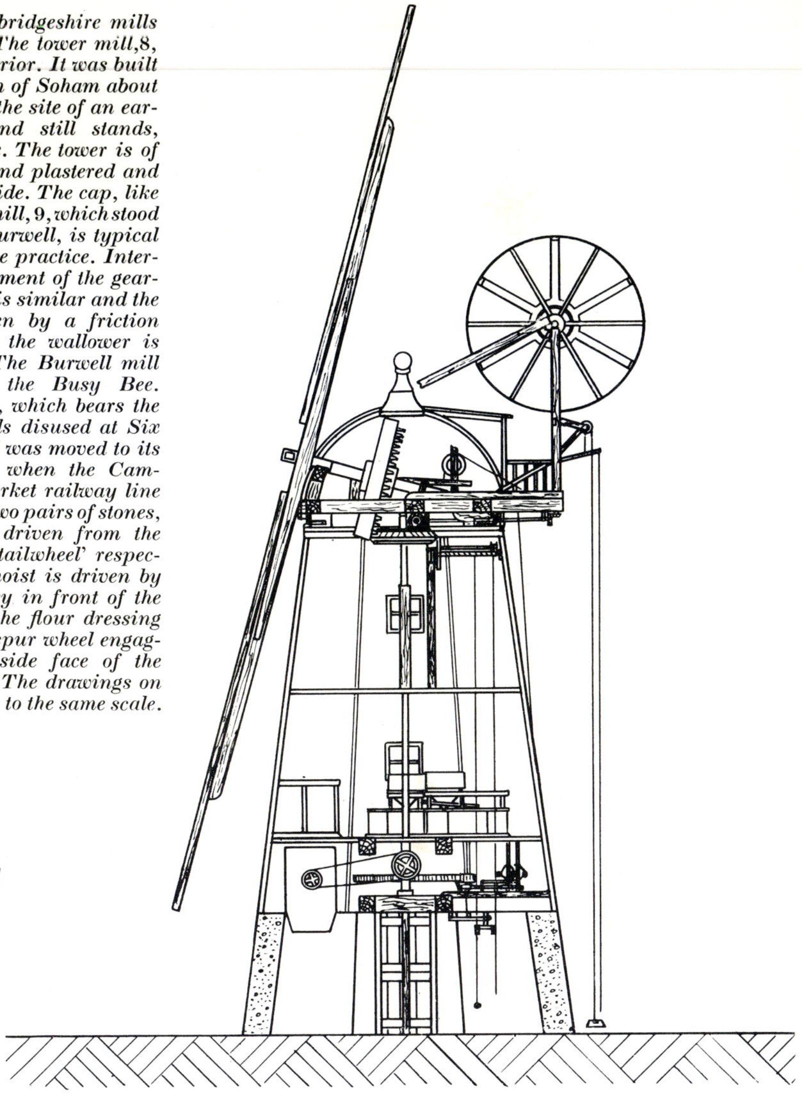

9

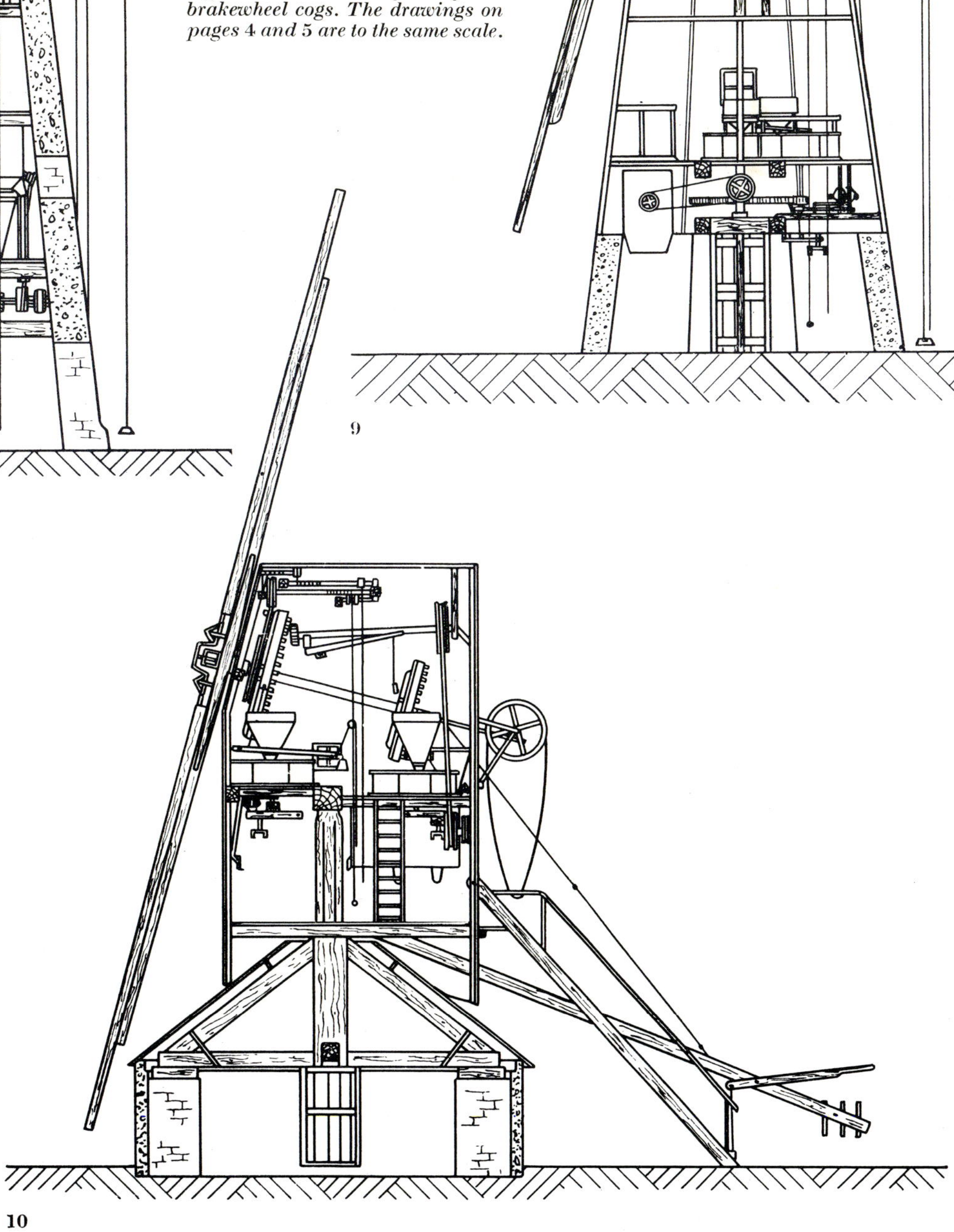

10

independent research had confirmed that it was in existence in 1636 the mill was bought and repaired by Mr. Alfred Bossom, M.P., and presented to the Cambridge Preservation Society. This was done out of funds raised by the exhibition of Epstein's *Genesis* in the house known as *Finella*, Cambridge.

Mills of a later date have curved roofs to accommodate larger gears and corn bins in the top. The oldest of this kind at work in this country is at Outwood in Surrey and was built in 1665, 101. It was given a pair of new sails in 1931 at the expense of the Windmill Section of the Society for the Protection of Ancient Buildings. A few years later I was able to present it with a couple of good second-hand sails from the post mill at Forncett End, Norfolk, which had been pulled down.

Porches give great character to a post mill, but many mills have none, 10 and 30. The simplest porch is no more than a narrow lean-to, 77, 78, 81, 99, some have pent roofs, 11, 65, and 66, some have a flat curve, others are bonnet shaped, 22 and 82. Perhaps the most attractive were some to be found in Norfolk and West Suffolk extending right across the tail of the mill and having a balcony of the same width, 79.

The internal economy of the early post mills is simple. A geared wheel is mounted on the windshaft and is known as the 'brake wheel' from the fact that a contracting brake to stop the mill acts on its rim, 15. The brake-wheel drives the 'wallower', and this drives a pair of stones direct. Grain from the hopper passes between the stones, is ground and passes down to the floor below as meal. Exactly the same type of power transmission was used in early watermills, with the difference that the power was transmitted upwards instead of downwards, a sluice-gate taking the place of a brake. It seems probable that the post mill is an adaptation of the watermill to circumstances where wind but no waterpower was available. Although none now remains in England, a variant of the post mill, known as the 'sunk post mill', existed at one time. Here the whole of the substructure, comprising the post, cross-trees, and quarter-bars, was buried in the ground, presumably for the sake of greater stability. The remains of such a mill have been found here and there, the latest being at Sandon Mount, Hertfordshire.

**the tower mill**

The effort required to turn the body of a post mill round is considerable, and the next step was the introduction of the 'tower mill', 1, 3, 4, 6, 23 and 29, also pages 4 and 5. This consists of a fixed tower, which enables more floor space to be provided for machinery and grain storage. A small top or 'cap' carries the sails, windshaft and brake wheel only, and is turned to face the sails into the 'eye of the wind'. The top floor under the cap is known as the 'dust floor', and others are known as the 'bin floor', 'stone floor' and 'meal floor', self-explanatory terms. The origin and date of this improvement is unknown, but the earliest illustration of tower mills dates from the middle of the fourteenth century. In England the earliest are on a fifteenth century quarry, 6, in Stoke-by-Clare Church, Suffolk, and the early sixteenth century carving, 4, on St. Mary Magdalene Church, Launceston, Cornwall, which have been mentioned before. The earliest reference to a tower mill in England seems to be a rhyme probably of sixteenth century origin.

'Glatton round hill,
Yaxley stone mill,
And Whittlesey Mere
Are the three wonders of Huntingdonshire.'

15

14

# A POST MILL AND ITS MACHINERY

*11, Worlingworth Mill is a characteristic East Suffolk post mill. The wooden 'buck' or body contains the machinery and the brick roundhouse below protects the substructure, 12, which here is viewed from the ground looking upwards. The 'post' can be seen quartered over the 'cross-trees' supported by the four 'quarter-bars'. The post, 13, extends through the lower floor of the buck and sockets into the 'crowntree' which is braced to the timbers below the floor with iron bars; behind it are the meal bins. On the floor above are the stones, 14, overdriven through the 'stone nuts', here out of gear, by a 'great spur wheel' mounted on an upright shaft. Above it is the 'wallower'. Grain in the bins on the top floor, 15, is fed through spouts into hoppers above the stones, warning being given by a 'bell alarm' if it runs low. The brake wheel, mounted on the 'windshaft' in the top of the mill, drives the wallower and a skew-gear for the drive to the flour dressing machine, while the 'sack-hoist, is driven by belt from the windshaft. To show a complete mill the photograph of the brake wheel, 15, is from Parham Mill, while that of the stones, 14, is from Saxtead Green Mill.*

12

13

*THE POST MILL at Great Chishall, Cambridgeshire, built in 1819 with material from an earlier mill, was fitted with a new post about seventy years ago. The oak came from Brandon in Suffolk.*

16

17

*THE TOWER MILL at Debenham, Suffolk, is typical in outward appearance of such mills built by the once famous firm of millwrights, Whitmore and Binyon, of Wickham Market.*

18, 19

20, 21

## TYPES OF WINDMILL CAPS

*Sutton Mill, near Stalham, 18, has a typical Norfolk boat-shaped cap. The gallery round the cap is convenient for repairs but bad for airflow. Smock mills in the neighbourhood of this one, 21, at Newmarket, had a characteristic cap and were probably the product of one firm of millwrights. The cap of the six-sailed Myer's Mill, Alford, 20, is typical of Lincolnshire, Yorkshire and the North Midlands. Kentish smock mill caps resemble the tops of post-mills like this one, 19, at Herne.*

Of surviving early tower mills one on the roadside between Norman Cross and Yaxley was pulled down within the last twenty years; while one on Sark, Channel Islands, also bears a sixteenth century date. In England, one of the oldest is at Chesterton in Warwickshire, 70, commissioned by Sir Edward Peto as an observatory in 1632, The conversion to a windmill probably took place during the early eighteenth century, and the structure is thought to have been the work of a pupil of Inigo Jones.

Tower mills built of timber and weatherboarded are usually octagonal, but may be six, ten or twelve sided, and are known as 'smock mills', their tapering towers and many sides resembling the old-fashioned countryman's smock. They are as a rule built on brick bases which may be anything from a few feet to a couple of stories high. The corner posts, suitably braced and tied, extend upwards, converging to give a considerable taper or batter to the tower, 83. To the top is fixed a circular iron-faced track or 'curb', built up of timber and on this well-greased curb the cap turns. The cap is built up on top of a heavy horizontal timber frame, below which blocks or rollers are fixed to run on the curb. The cap is usually centred by means of 'truck wheels' on vertical spindles attached to the cap frame and running against a vertical track on the inner side of the curb. In the north-west, however, a flanged iron ring was sometimes fixed to the top or dust floor of the mill and the cap was centred by means of a device resembling an inverted table with a hole in the centre to fit over the ring, the 'legs' being cross-braced to each other and attached to the cap frame.

In some cases in East Anglia the cap turns on a 'shot-curb'. An iron track is fixed to the bottom of the cap frame as well as the top of the curb, and between the two tracks runs a set of rollers, closely spaced and held in place by two concentric rings in which the roller ends bear. It is in effect a roller thrust-bearing, and can be compared to the ball-bearing collar on the post of Sprowston Mill, Norwich.

Towers of timber are very vulnerable to the weather, which gets in at the corners and rots the corner posts. Round brick towers are now more common in this country and, if the windows on successive floors are spiralled round the tower and not placed one immediately above the next, they will stand for many years without need of structural repair. Sometimes, however, foundations are not sound and the tower eventually leans, and in some cases has been jacked upright and underpinned. A distorted tower means a distorted curb; the cap will not turn and the sails are damaged when they are 'tail-winded', i.e., when the wind takes them from the back. In Anglesey, 29 and 85, and the north-west, towers were often of stone, while towers of clunch and clay lump have been built in Eastern Counties.[1]

Caps are the distinguishing feature of tower mills, each area having its own traditional designs, 1, 7, 8, 9, 17, 18, 19, 20 and 21, though now that so many mills have disappeared it is not always easy to tell what was the tradition for a given area. The most characteristic are the Kentish wagon shape, 105 to 108, the boat shape of Norfolk, 18, and the north-west, 29, 85, and the ogee of Lincolnshire, 20, 23, 91 and 92, and the north-east. There are of course many variations, some even traditional to a small area, such as the Newmarket type, 21, of pepper-pot cap with an acorn-like finial. The Lincolnshire ogee cap with its ball finial suffers from the great disadvantage that in many cases it does not cover the front and rear of the cap frame, which is thus liable to suffer from the weather. The very rounded boat-shape Norfolk cap, an early nineteenth century innovation, is probably the best design from the point of view of construction, protection, and aerodynamic efficiency. Both the cap and the tower of a mill may sometimes have a stage built round them, 17, 18, 23, 83, 84, 89, 91, 92, 94, and 103 to 108. In the case of the cap it is used for ease of access to the outside and to the sails; in the case of the tower it is generally at the level of the floor on which the stones are situated, and is used for the adjustment of the sails. The stage round the tower usually improves the appearance; but that round the cap may spoil its lines and certainly spoils the slipstream behind the sails.

**the composite mill**

A few examples of the 'composite mill', 68, existed in the Eastern Counties, although none now remains at work. In each case the mill consisted of a post mill body which had been removed from its substructure and mounted on a short tower, having a curb at the top, on which it turned. While the body of the composite mill at Banham, in Norfolk, was being moved from

[1] A clunch-built mill at Waltham, near Grimsby, Lincolnshire, was undergoing repairs during a spell of very wet weather. One afternoon, after having had lunch in the miller's house, the millwrights went out to find that the walls of the mill had quietly collapsed and the mill had sunk down upon itself.

22

*COUNTRY MILL. Wetheringsett post mill, Suffolk. The mill was blown down in 1881 and rebuilt.*

23

*TOWN MILL. The five-sailed Maud Foster mill at Skirbeck, Boston, is the last of the mills in this town to work by wind. It takes its name from the Maud Foster drain in the foreground.*

elsewhere the wagon on which it was being transported became bogged. It had not reached its destination at the top of the ridge, but the tower was built there and the mill body mounted on it.

**moving windmills**

A mill was always considered a portable object and there are numerous instances of post mills being moved to new sites. Perhaps the best known is one celebrated in a contemporary print which was a post mill moved from the site of Regency Square, Brighton, in 1797. The sails and machinery were removed, the mill mounted on rollers, and eighty-six oxen loaned by neighbouring farmers with their steady pull moved the mill slowly but surely to its new position, near Withdean, a distance of over a mile. Smock mills, like post mills, were also regarded as portable property, The Nettlebed mill, burned down in 1912, originally stood below Watlington, Oxfordshire. In 1823 or 1824 the machinery was removed and the mill was hauled on two timber carriages by sixteen horses, At the top of the hill both carriages broke down, and the mill rolled over and over, and stopped just at the place they wanted to fix it.

Of tower mills that have been moved Heckington, 45, is the best known. Standing on what is now the site of Boston Docks, Lincolnshire, the mill was bought, dismantled and re-erected in a burnt-out tower at Heckington near Sleaford, the owner doing the work himself with the aid of direct labour.

**winding the mill**

To get the most power from a mill it is necessary that the sails should always face square into the 'wind's eye'. This 'winding' of the mill was first accomplished by pushing the whole body of a post mill round by means of the 'tail pole', which projected downwards through the ladder, 26. Normally the ladder rested on the ground and acted as a back stay while the mill was at work. Before turning the mill the ladder was raised off the ground by depressing a lever pivoted on the tail pole and connected to the ladder by a chain. The lever could be held down—and the ladder held up—by means of an iron pin, while the miller pushed against the tail pole either with his back or with his shoulders against a yoke which projected below it at its extremity. In the case of the early tower mill caps a number of posts was set round the mill, a portable winch was anchored to one of them, a chain run out to the tail pole and the cap was drawn towards the winch by winding up the chain. An improvement on this was to fix the winch to the tail pole, 67, and later, to fix a geared rack to the curb at the top of the mill, 29, 85. A chain wheel at the side or back of the cap was connected to the rack through gearing and an endless chain hung down from it to the ground or to a stage round the mill. By pulling this chain the cap could of course be turned. A method of turning the mill into the wind automatically was patented in 1745 by Edmund Lee. In the case of a tower mill a wind wheel of from five to eight vanes, 28, is set at right angles to the sails at the tail of the cap and connected through gearing to the rack on the curb. So long as the sails face squarely into the wind this automatic 'fantail' remains stationary; but when the wind veers it strikes the sides of the vanes and revolves the fantail, which turns the cap square into the wind once more. This fantail is also applied to post mills and is usually mounted on the ladder, 11, 16, 22, 27, or the tail pole, 102, in which case it drives wheels in a carriage running on a track round the mill. Occasionally it is mounted on the top of the mill, 30, 79, in which case the drive is sometimes

27

24

25

26

*24, Keysoe Mill, Bedfordshire, at rest, the tail pole supported in a crutch and a ladder rests on the ground. 25, The miller raises the ladder by means of the lever. 26, He pushes the mill round into the "eye of the wind" by means of the tail pole. 27, Fantail and fan carriage, Friston post mill, Suffolk, 28, Fantail, Wellingore tower mill, Lincolnshire. Each arm is socketed into a cast iron hub, two wrought iron rings connect and stiffen them and a balancing weight can be seen on one blade.*

28

taken vertically downwards, and then in at right angles to a worm wheel mounted on the post.

**windmill sails**

The earliest mill sails, 31 to 35, were simple wooden frameworks consisting of a number of evenly spaced horizontal bars mortised through a stout back-bone or 'whip', and held together on either side by a 'hemlath' running the full length of the sail. Two stouter timbers called 'stocks' were mortised at right angles through the end of the windshaft and wedged in place, and to each a couple of sails were bolted and strapped. Sail-cloths, laced in and out of the sail bars, could be set by being drawn outwards from the whips to the hemlaths and the sails had a constant angle of weather.

29

*Cemaes Bay Mill, Anglesey, 29, is typical of Anglesey, and to some extent of the mills of the North-West in general. The weather of the sails is so pronounced that the lower sail appears to be acutely tapered. The cap is turned by the endless chain hung from the wheel at the rear. To prevent it swinging a chain is hitched to one of the wooden cleats on the mill tower. The stonework platform prevents people or animals approaching the sails too closely. Anglesey was the granary of Wales and at least 35 sites of windmills can be found. None now remains at work, 85, the last to work being Stanley Mill, Tre-Arddur Bay, now destroyed, see also 43. Common sails were universal.*

30

*As in the case of other counties, windmills in Essex have suffered an eclipse until there were but two at work where once there were over one hundred. Of these the post mill at Ramsey, 30, was one. Built about one hundred years ago by Whitmore & Binyon, the Suffolk millwrights, this mill was owned by Scotts, the biscuit makers of Ipswich, and before the war was grinding for them. The mill is in the Suffolk tradition. Post mills in Essex usually had lower round-houses and were turned by a tail pole. Ramsey mill is no longer working.*

Later, sails were improved by being given 'weather'—a twist, 43 to 46, something like that of a propeller—and sail area only on the driving side of the whip; while a 'leading board' inclined at an angle—ideally $22\frac{1}{2}°$—on the driving side helped to hold flat the sailcloth now spread over a framework and not laced in and out of it. This is known as the 'common sail' and each millwright had his own design arrived at empirically, John Smeaton, of lighthouse fame, being the first to investigate sail design on a scientific basis.

In 1772 Andrew Meikle, who was credited with the invention of the fantail, introduced the 'spring sail', 37 to 40, 78, 87, 93, 101, 103. For canvas he substituted a number of small hinged shutters, like those of a venetian blind. All the shutters in a sail are connected together and are controlled by a spring whose tension can be varied according to the power required. Should the wind pressure increase beyond the required point the shutters hinge open and 'spill the wind'. The spring of each sail has to be tensioned separately, but this done the regulation is automatic. A further step forward was the invention, by Captain Stephen Hooper, in 1789 of the 'roller reefing sail', 42, 86. In place of shutters are substituted small roller blinds, the bottom of one blind being attached to the next by webbing straps or 'listings'. All can be operated simultaneously by two 'airpoles' to each sail, which are connected in front of the windshaft through cranks and levers to a spider coupling. The windshaft is drilled right through the centre[1] and a 'striking rod' passed through it. The rod is connected at the front to the spider coupling and at the back to a chain wheel by means of a rack on the end of the striking rod engaging with a pinion on the chain wheel spindle. By pulling the endless chain which hangs over the chain wheel down to the ground the striking rod can be moved backwards and forwards like the boss carrying the ribs of an umbrella. The blinds open or close at will; but the method is not automatic and although all the sails can be set at the same time without stopping the mill the blinds and their listings are more liable to damage than the shutters of a spring sail, which are of wood or of painted canvas on wire or wooden frames.

In 1807 Sir William Cubitt invented what has ever since been called 'patent sail', 46. He combined the shutters of the spring sail with the remote control of the roller reefing sail, and weights, 47, are hung on one side or the other of the endless chain according to whether the shutters are to be kept open or closed, 1. The mass of the weights is altered according to the different velocities at which it is desired to 'spill the wind', and all this can be accomplished on all the sails simultaneously without stopping the mill. In place of a rack on the end of the striking rod a sliding block is often used, and to it is attached a rocking lever projecting from the tail of the cap. From this a chain is hung down and then looped up over a pulley fixed to a part of the fan supports and back to the top of the lever. In this way it is possible to move the lever up and down by pulling on one side or the other of the looped chain, and the setting of the shutters is done in the manner already described. Patent sails can be single, like common sails, or double, like the early cloth spread sails; but the single shuttered sail is probably the most efficient although not the most usual.

In 1769 John Smeaton introduced cast iron into millwork. Wooden shafts were replaced by iron ones one-third of the former diameter, 48, and usually an iron 'poll end' is cast at the front end of the shaft. This resembles two boxes at right angles without any ends and through them the sail stocks are passed and wedged or pinched into position. In the north-east and north-

[1] This was done while the mill was at work, the drill being lengthened 18 ins. at a time.

31

35

32

36

33

37

34

38

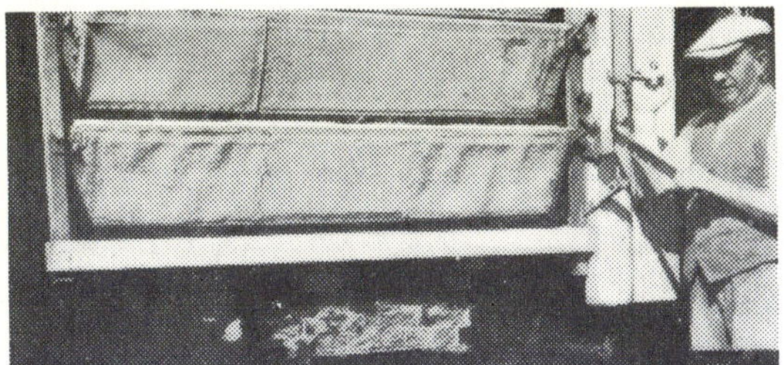
39

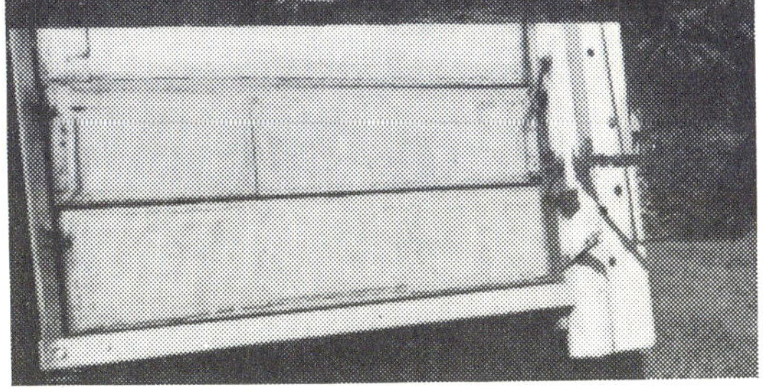
40

41

## SETTING AND REEFING SAILS

*Keysoe Mill, Bedfordshire,* 41, *was rebuilt in* 1801 *and blown down in* 1945. *It has two common and two spring sails.* 31, *the cloth is wound up and wrapped round the frame of the common sail.* 32, *the miller unrolls the cloth.* 33, *he spreads it to 'sword point' by means of one of the 'pointing lines'.* 34, *he spreads it further to 'first reef'.* 35, *it is spread to 'full sail'.* 36, *the miller releases the brake by means of the rope hanging out of the side of the mill to bring down the next sail for adjustment.* 37, *a spring sail, with the shutters open. The miller is about to adjust the tension of the spring.* 38, *the shutters closed.* 39 *and* 40, *the adjusting mechanism.*

4

43,

## WINDMILL SAILS COMPARED

*The hinged shutters in the leading boards of the 'roller reefing sails' on Tollerton Mill, Yorkshire, 42, act as air brakes. The common sail, 43, on the Melin y Bont, Bryn-du, Anglesey, is typical of those on mills found in the North-West. The apparent taper is the effect of the pronounced 'weather'. The inclination of the mortises for the 'bars' should be noted, as this is the method by which the weather of the sails is obtained. The patent sail, 44, is on the smock mill at Dalham on the Suffolk-Cambridgeshire border. The mill is now preserved as a landmark and the shutters removed for safety. The 'bays' into which the sail is divided are clearly shown and through one the derelict smock mill of Ashley can be seen just across the border in Cambridgeshire. The patent sails on Stretham, Cambridgeshire, 46, are shown in good order complete with their striking-gear; 'clamps' are fitted to the front pair. The eight-sailed mill at Heckington, Lincolnshire, 45, originally stood at Boston and was bought and transferred to an empty tower by the late owner when Boston docks were constructed.*

45, 4

*Nesting weights for the striking chains of the patent sails at Wellingore Mill, Lincolnshire.*

48

49

50

## WINDSHAFTS

*The 'poll end' cast solid with the windshaft is the most usual mounting for sails and in this picture of one taken out of Great Mill, Wickhambrook, Suffolk, 48, the neck journal can be seen behind the poll end and behind that the ribs on which the brakewheel was hung. In Lincolnshire and the North a cast iron 'cross' is hung to the front of the windshaft with keys instead of a poll end as can be seen, 50, in one at Swineshead North End Mill, Lincolnshire. The arms of the cross are of Tee section and the sails are bolted and strapped to them. The wooden clamps on the front of the sail 'backs' prevent the straps from cutting in. An open bearing and a plain 'brass', is used for the neck journals as at Heckington, 49. In the old days these brasses were often cast from melted down clock faces, candlesticks and pestles and mortars.*

west, however, the front of the shaft is plain, and a cast iron 'cross', 49, 50, is hung on to it with keys. Stocks are dispensed with and the sail whips, somewhat stouter than those used in the South and now termed 'backs', are bolted and strapped to the arms of the cross. This method obviates the use of heavy and costly timber stocks which, if they break, will throw what may be a sound sail and necessitate the removal of the opposite sail when the stock is renewed. With the cross the removal and refitting of a sail is considerably simplified, and by increasing the number of arms the number of sails can be easily increased, which is not the case with a poll end. So in Lincolnshire, where the cross is almost universal, five- six- and eight-sailed mills, 20, 23, 45, 86, 91, 92, were built and a few still remain. John Smeaton built the first five-sailed mill at Newcastle-on-Tyne, and this number of sails is probably the most efficient. But a six-sailed mill is far more convenient, as, if one sail of a five-sailer is damaged the whole mill is out of action until it is replaced, whereas a six-sailer can be run with four, three, or two sails if necessary. Eight single-sided sails is probably the limit to the number that can be usefully employed, as a windmill can be overcrowded with sail just as can a sailing-ship. Four mills have also been built with annular sails; the best known of these was at Haverhill, Suffolk, 69, built about 1860. It had a sail 50 ft. diameter with 120 shutters and about the same power output as an ordinary four-sailed mill.

It should be noted that all windshafts carrying the sails are inclined at an angle to the horizontal of from ten to fifteen degrees. This is primarily to throw some of the weight back to a thrust bearing at the tail of the shaft inside the mill, and hence to balance the weight about the 'neck' or front bearing. It also enables the sails to clear a tapering tower or the roundhouse of a post mill.

**windmill gearing**

While gearing to-day is a highly specialised form of engineering, few modern engineers are acquainted with the work of early millwrights in this direction. The windmill, since it makes use of an uncontrollable source of power, presents more interesting applications of early gearing than any other mechanism. The earliest windmills, like the earliest watermills, possessed no gearing; but this was probably introduced into millwork with the invention of the Roman watermill, described by Vitruvius. It required a change in the direction of the drive, and the gearing of the post mill is substantially the same as that of the Roman watermill, but with the drive coming down from above instead of up from below.

In a windmill the first gear is the 'brake wheel', which is mounted on the windshaft to which the sails are fixed. Early windshafts were of wood and the brake wheels were of the 'compass-arm' type, with arms mortised right through the windshaft, weakening it where it needed strength. Later came the 'clasp-arm' wheel, 54, with four arms forming a square at the centre through which the windshaft passed. All these wheels were of wood and very massive, the rims built up of cants and strakes and with heavy wooden contracting brakes, 51, acting on their rims. Then when John Smeaton introduced cast-iron into mill work iron windshafts, iron wheels, or wheels with iron hubs and spokes and wooden rims, or in the case of brake wheels, iron cogs cast in segments and bolted on to the rims, 51, all came into use. The first driven wheel in a mill is called the 'wallower' and the first wallowers were wooden 'lantern pinions' with staves instead of cogs fixed between top and bottom flanges and mounted on a vertical shaft. From these developed the 'trundlewheel', in which the top flange was dispensed with and the staves shortened. Finally the wallower was built as a bevel wheel either of wood or of iron or of both. In post mills with stones fore and aft two wallowers are required, and the stones are driven direct from the brake wheel, 10, in front and the 'tail wheel', also mounted on the windshaft, at the rear. When the stones are mounted side by side in the breast of a post mill, 14, and in the case of tower mills, 8, 9, the drive is taken through the wallower at the top of the 'upright shaft' to the 'great spur wheel' at the bottom of the upright shaft, which drives the stones through pinions known as 'stone nuts'. To get the increase of speed required the great spur wheel is often very large and varies in construction as much as the other gears.

The first stone nuts were lantern pinions, spur gearing being a later development. When the stones are overdriven the nuts are mounted on vertical spindles or 'quants', 14, which drive the 'runner stones' from above; when underdriven the stone nuts are mounted below the 'bed stones' on the 'stone spindles', page 5, on which the runner stones are balanced. To disengage the stone nuts from the great spur wheel in the case of 'overdrift' stones the quant is released from its top bearing and allowed to fall sideways. In the case of 'underdrift' stones, however, the nut is either lifted out of gear with the great spur wheel or else several 'slip-cogs' are removed.

Drives for auxiliary machinery are usually taken by a spur wheel or a skew gear from the brake wheel, 15, in the case of post mills and from a geared bevel fixed to the great spur wheel or independently to the upright shaft in the case of tower mills. Sack hoists are similarly driven, page 5, but in addition belt drive from the windshaft is used in the case of post mills, 15, and friction drive from a friction ring below the wallower in the case of tower mills, 51.

Gearing is also used to turn the sails of the mill into the eye of the wind, and the invention of the fantail was made possible by the use of cast-iron gearing, a number of small bevel and spur gears being used on both tower and post mills. The only use of the worm is found in connection with the winding gears, the earliest worms and the cogs of the racks they engaged being of

31

52

53

54

## DETAILS OF GEARING

*Caps house the windshaft on which the 'brakewheel', 51, is mounted. A large wooden contracting brake acts on the rim of this wheel at Wymondham Mill, Norfolk, which drives the 'wallower'. The wooden friction ring below the wallower is for driving the 'sack hoist'. The Vee-shaped cuts in the 'cants' forming part of the rim of the brakewheel show that it has come from elsewhere. It probably had wooden spokes instead of the present 'clasp arms'. and was mounted on a large wooden windshaft through which these spokes were mortised. Wooden shafts were commonly three times the diameter of the iron ones which replaced them. In Pakenham Mill, Suffolk, 1, the 'great spur wheel', 53, mounted on the bottom of the upright shaft, has iron arms and a heavy wooden rim. In addition to the main set of wooden cogs driving the stone nuts there is an iron internal gear inside the rim driving one pair of stones. The underside of the iron wallower of Denver Mill, Norfolk, 52, to which is fixed a bevel ring. This drives the sack hoist, which can be seen behind and slightly to the left of the upright shaft. The brake-wheel, 54, of Laughterton post mill, Lincolnshire, drives a pair of stones direct. It is mounted on a large wooden windshaft and has an iron brake.*

*The roof construction of a cap varies according to its shape. An exterior view of this one at Diss, 55, is to be seen in 95.*

55

*Peak Bedstone at Union Mill, Cranbrook, Kent, 56. It has been newly dressed and on it are the quant and mace which drive the runner stone. Wire machine at Keysoe Mill, Bedfordshire, 57. Stones at Heckington Mill, Lincolnshire, 58.*

56, 57, 58

wood. The most usual timber used for cogs is hornbeam, which is both tough and porous; certain apple wood is better but nowadays is almost unobtainable.[1] To prepare a cog the blanks are first sawn from the plank and placed in a wooden jig or 'cog box' in which they are shanked, the shanks are shaped and fitted to the mortise gear, the face and sides turned in a lathe and then pitched out and trimmed by hand, usually to an epicycloidal shape. They are held in place by two main methods; either the shanks are drilled and pins are driven in, or the shanks are dovetailed and wedges driven in between them. A wood to wood drive using well-shaped gears correctly lubricated is very quiet and will last for many years; but if speeds are high, wood to iron is more usual.

**stones**

Grain is usually stored in bins above the stone floor and is fed through a spout to a hopper, 58, also page 5, supported on the stone casing by a wooden frame or 'horse'. From there it passes down an inclined 'shoe' held at its upper end freely and at its lower end by a spring against the quant, in the case of overdrift stones, or in the case of underdrift stones, against a three- or four-cornered removable extension of the stone spindle called a 'damsel',[2] As the quant or damsel rotates it vibrates the shoe, the grain is shaken down and grain flow can be controlled by a 'gate' in the shoe or by altering its angle of inclination. The grain falls into the 'eye' of the runner stone, is caught by the 'furrows' and as it is ground passes outwards to the edge. It escapes as meal into the space between the stones and their casing and falls down a spout to the meal floor below, 13.

A piece of leather is fixed low down inside the hopper and to it is attached a string which passes through the side of the hopper to a bell. So long as there is grain in the hopper, it presses on the leather and keeps the string taut. When the weight of the bell overcomes the weight of the grain on the leather the bell falls by its own weight against some moving part of the machinery and warns the miller that the grain is running low. The gap between the stones can be altered to ensure a constant degree of fineness of grinding irrespective of the speed of the mill. This is effected by raising and lowering the 'bridgetree', on which the stone spindle carrying the runner stone is mounted, by means of a system of levers. In early days this was done by hand; later, centrifugal governors were used to carry out this 'tentering' automatically.

Two kinds of stones are in general use: the Derbyshire 'peak stone', a millstone grit taking its name from the Peak District where it is quarried in one piece from the living rock, and the 'French burr' stone, a freshwater quartz much harder than the peak stone. It is quarried at La Ferté-sous-Jouarre in the Paris basin in small blocks, which are shaped like the stones of an arch, built up to the required diameter and cemented together with plaster of Paris. 'Blue stones' or 'cullin stones' are used nowadays only in the North for barley, oats and rye. They are a lava of a bluish-grey colour once sold from Cologne (hence their name) and quarried at Neider Mending in the Rhineland. Local stones were also used in places though not to any great extent, and nowadays composition stones of emery and cement are widely used.

**machines**

After being ground, wheat meal intended for conversion to flour was

[1] Holly, beech, evergreen, oak and other woods were also used.
[2] So called because it makes more noise than anything else in the mill.

## MILLERS

59, 60, 61

*59, Mr. Elmer, of Woolpit Mill, Suffolk. 60, Mr. Davey, of Legbourne Mill, Lincolnshire. 61, The late Mr. Aldred, of Saxtead Green Mill, Suffolk.*

## MILLWRIGHTS

62, 63, 64

*62, The late Mr. Robert Martin, senior, of Beccles, Suffolk. 63, Mr. T. B. Hunt, of Soham, Cambridgeshire. 64, The late Mr John Bryant, of Colchester, Essex.*

passed through a bolting machine or 'bolter' to dress off the bran. A few of these machines are still to be seen though none is in operation to-day in windmills. They consist of an inclined wooden shaft to which radial arms are fixed. These arms carry light bars running parallel to the shaft, they form the reel and just clear some stout beaters fixed to the casing of the machine. A seamless woollen bolting cloth, rather larger in diameter than the reel, was tied to it top and bottom. The reel revolved and the cloth sagged outwards with the weight of the meal admitted at the upper end. It hit the fixed beaters and the flour passed through the cloth into the casing, leaving the bran to tail out at the lower end.

Another method was to use a wire machine, **57**. This consists of an inclined fixed wooden skeleton cylinder lined with wire mesh of varying degrees of fineness. Down the centre of the cylinder passes a wooden shaft carrying several rows of longitudinal brushes. As the shaft revolves the brushes force the flour and the middlings through the wire mesh into the casing and the bran tails out at the end. Another machine found in the north-west was the groat machine. Oats were roasted in a kiln and passed between a pair of stones with a gap between them equal to the length of an average oat. This split the husks, which, with the kernels or groats, passed down into the groat machine. Here they were put through a wire machine which brushed out the red dust produced in the roasting and as they tailed out they were subjected to a blast of air from a four-bladed wooden blower or fan. The heaviest groats fell down into a nearby spout, the lighter ones into a spout further on and the husks were blown through into the 'husk cupboard'.

### millers and millwrights

What of the builders of the windmills, these millwrights, the ancestors-in-trade of the present-day mechanical engineer? They are for the most part unknown, 76, they worked in small country places, 62, 64, and most of them, like their monuments the windmills, have gone without leaving a trace. Some few still remain, 63, their businesses established anything up to two hundred years ago. Some have left the millwrighting business and become production engineers, and a number who did this a hundred years or more ago failed to survive when the centres of industry shifted or foreign markets were closed to them. The ingenuity of the millwrights is constantly shown not only in the design of windmills, arrived at empirically over a long period of years, but also in the smaller details such as bell-alarm and sack-hoist operation, the variations in the striking gear of patent sails and 'gadgets' such as the sack loader, 83. They left few records and drawings; only a small proportion of their work survives and it should be recorded before it is too late, for it is work to be proud of, ingenious, honest and expressive of the best tradition of their craft.

Those who work and repair windmills, 59 to 64, are usually men of considerable force of character with a capacity for hard work. They had to be tough physically, too, and usually were as the following will show. At Billingford, Norfolk, a post mill once stood where the tower mill now stands. In the sixties it was tail-winded and blown over forwards. The miller was in the breast of the mill at the time on the lower floor, and although he fell with the mill the post protected him from fatal injury. His man was not so lucky; coming up the ladder at the back he was thrown into the debris and died of his injuries. The late Mr. John Bryant, 64, millwright, of Colchester, who died in February, 1944, was working on the sails of Great Holland Mill, Essex, when the miller closed the shutters, trapping his fingers, and took off the brake. The sails started to turn and the ladder and tools, falling with a clatter on the bakehouse roof, brought the baker running out. He stopped the mill when a revolution and a half had been made leaving the sail, on which the millwright was trapped, in the top position and the millwright, of course, upside down. After the sail had been got to the bottom position Mr. Bryant was helped down but he could not straighten his fingers and had to drive sixteen miles home in his trap with the reins round his arms.

### marsh mills

Two only of the typical large marsh mills of the nineteenth century were in use until quite recently; that at Soham Mere, 98, which still retains two of

[continued on page 32]

65

## THE MECHANISM OF A POST MILL

*65, 66, a true to scale model 1/12 full size, of Sprowston Mill, Norwich, 1780–1934, made by the late Mr. H. O. Clark of Norwich, who started work on it in June 1924 and completed it just fifteen years later. It is similar to the mill shown in Crome's painting of Mousehold Heath in the National Gallery, though the round house, fantail and patent sails were modernisations. A fund was raised by Mr. Clark and some of his friends to repair the mill as a landmark, but a few days before the work was due to begin the mill was burnt to the ground.*

*The design of the mill is representative of the best East Anglian post mills, that is to say the best in the world and may be summarised as follows. There are four double-shuttered patent sails with eight bays of three shutters each side. The angle of the sail bars varies from a few degrees at the tip to 27 degrees at the centre. The striking gear is operated from inside, the striking chain passing down through the stone and meal floors into the roundhouse for convenience of operation. The brakewheel is of the clasp-arm type, made of wood, with a wooden brake acting on its rim. It drives a metal wallower mounted on a metal upright shaft and the two pairs of stones in octagonal vats in the breast are underdriven from the great spur wheel with metal teeth by stone nuts with wooden teeth. They are fed from grain bins on the top floor and feed into meal bins on the bottom floor of the mill body. The flour bolter across the tail of the mill is fed by a separate bin and discharges into sacks by three stockings. The roundhouse is of brick with a boarded roof, and the six-bladed fantail drives the carriage wheels through twelve gears.*

*There are four unusual features in the design of the mill. The first is a cone clutch driven from the brakewheel to operate the sack hoist. The second is a shaft drive from a bevel rim on the great spur wheel to a pair of stones in the tail of the mill. The third is the cambered side girts; one is clearly shown in*

66

*the detail illustration. The fourth is a ball bearing for the post. This is a steady bearing just below the bottom floor of the mill body. The original balls were of cast iron, 52 in number and $2\frac{1}{4}''$ diameter and were possibly coeval with the mill.*

*The flour bolter in the tail of the stone floor had long been removed, but one in Freckenham Mill, Suffolk, was copied and the model cloth for it was from another full-sized bolter at Haddenham, Cambridgeshire. The model stones are made from steel, correctly dressed, hardened and ground, as nothing less would carry out actual grinding on this scale, the models weighing but $1\frac{1}{2}$ lbs. against the 2600 lbs. full size. It is satisfactory to know that this superb model is now in the Science Museum, London, a fitting memorial to its builder.*

its sails, and the Betty Mill, Nordelph, demolished now. At the latter the marsh man and his family actually had their living and sleeping quarters in the body of the mill. The smoke from their fire escaped into the mill and found its way out as best it could. In gusty weather the family would retire to the adjoining engine house to escape the smoke blown back into the room.

Marsh mills, 67, also page 4, used for draining the land drove large wooden 'scoop wheels', like paddle wheels, in a close-fitting casing, which raised the water a height equal to about two-fifths of their diameter. The drive was effected through a pair of wheels at the base of the upright shaft, the driven wheel being mounted on the scoop wheel shaft.

These mills in the Fens have from time to time been converted into grist mills, and examples could at one time be seen at Christchurch in Cambridgeshire, Dyke in Lincolnshire, Freckenham in Suffolk, and Salter's Lode in Norfolk, to mention a few. All were smock mills, and in the case of Dyke there was a tradition that it was moved from a neighbouring fen 'in Boney's time', while Freckenham first ground grist in 1801.

On the Norfolk Broads marsh mills were recently more numerous since the Broads, unlike the Fens, have not been dried out and sunk as they were drained; but electrically driven centrifugal pumps are replacing the mills as a result of the spread of the 'Grid', and it seems probable that even here the complete disappearance of the windmill as a prime mover is only a matter of time, for only four now work by wind.

**the future**

It is the low efficiency of the windmill that not only reduces its power output, but prevents it from working in light winds. In 1924, as a result of a competition held by De Hollandsche Molen (the Dutch Windmill Society), the Dutch millwright Dekker successfully overcame much of this trouble by inventing a special roller bearing which could be applied to the front or 'neck-journal' of the windshaft, and by applying aerofoils to the leading edges of the sails, 71. In 1925 a windmill fitted with the 'Dekker system' was given a year's trial by the Research Institute of the Government Aeronautical Service at The Hague. It was found that it gave three times the power output of a normally equipped mill, and that in the case of a Dekkerised mill in Amsterdam on only one day in the year was there insufficient wind to turn the mill. The Dekker system of streamlining was taken up with enthusiasm in the Netherlands and Belgium together with modifications by other millwrights; especially one by van Bussel which does not involve the complete enclosure of the whip. So that before the freeing of the Netherlands from the Nazis between three and four hundred mills had their sails streamlined in various ways. The work went on even during the Nazi occupation and in the case of some drainage mills the increased availability has made it possible to take out stand-by oil engines and rely on wind power alone.

Owing to their construction it is not possible to fit aerofoils to sails built on whips and fixed in pairs to stocks, as in England; and so far no satisfactory method has been worked out of applying aerofoils to the single shuttered patent sails mounted on an iron cross as in Lincolnshire. Should a solution be found it may be on the lines of a completely streamlined sail, folding backwards from each side of the whip as wind pressure increases, as designed and built by the late Mr. Lancaster Burne before the 1914–18 war.

In Holland no whips are used and the bars of the sails are mortised

67, 68

*67, a marsh mill at Tunstall, Norfolk, built in 1818. 68, a composite mill at Laver, Essex.*

69, 70

*69, the annular sailed mill at Haverhill, Suffolk. An unsuccessful attempt was made to preserve the mill as a landmark. It became derelict and when examined in 1942 the following lines were found pencilled on the whitewashed wall:*
*To the people of Haverhill,*
*who couldn't spare a pound or two*
*to keep this place in decent condition*
*Ye scunners a'*
*That couldna' thole*
*A place like this tae staun.*
*Whit's a bawbee?*
*Damned ane ye'll need*
*The place where ye're a' gaun.*
*70. A converted observatory at Chesterton, Warwickshire, dated 1632.*

directly into the stocks, which are held in poll ends as in England. The stocks are often built up of steel, and the method of construction and erection of a pair of streamlined sails in French Flanders in 1938 can clearly be seen, 72 to 75.

It seems possible that with the great advance in the technique of building aluminium rectangular tapered booms for aeroplanes it should be possible to adapt it to the needs of windmills particularly in view of the scarcity and high

71

*Dekkerised sails, Kouderkerke Mill, Isle of Walcheren. By streamlining sails and fitting roller bearings to the windshaft the output of mills such as these has been enormously increased. This Dekker system might be applied to English mills carrying sails mounted on a cross in the Lincolnshire fashion. In Holland the bars of the sail are mortised directly into the stocks which are often of steel, and no whips are used.*

72

*Dekkerising the Moulin de St. Arnould, Terdeghem, Steenvoorde, Nord. The stock is built up of steel plate and riveted, the mortises for the bars being lined with sheet metal, 72. Both the sails are built up on the ground. One is then dismantled and the stock drawn slowly through the poll-end of the windshaft, horses being harnessed to the rope tackle, 73, 75. The bars are fixed in place by the millwright during the process, 74. The ribs on the leading edge support the aerofoil.*

73

74, 75

***DEKKERISING***

cost of pitch pine, which is the only available timber suitable for sails under normal peacetime conditions. A cheap and compact method of storing electricity has yet to be found and it is likely that any such discovery will be too late for the salvation of the old-fashioned windmill in the role of a generator of electricity.

Two further possibilities are envisaged. A mutual insurance scheme for windmillers to cover storm damage to their mills. Damage to sails is the most common and the most expensive. Should a pair of sails have to be renewed the result is a three figure bill and only too often a bad storm spells the death of several windmills. As a corollary to this it will be necessary to initiate an energetic apprenticeship scheme to ensure that the race of millwrights does not die out with the present generation. At the present time a boy can not only obtain but really earn much higher wages on a farm than a millwright can economically afford to pay him while he is learning his trade. It will be necessary, therefore, to call in the help of existing apprenticeship charities which were founded to assist similar cases though in different trades. If the Society for the Protection of Ancient Buildings takes the lead in such matters it should have the fullest possible support from all who have an interest in the windmills of England.

76 *Tombstone 'to the memory of William Beeton Gent, formerly millwright and contractor, died in July,* 1822'.

78, 79, 80

**Suffolk.** *Nowhere in the world was the post mill brought to such a high pitch of perfection as in Suffolk in the nineteenth century, with patent sails, fantails and the most up-to-date gearing and equipment then available for windmills. The West Suffolk mills often had attractive flint and brick roundhouses as at Drinkstone,* 78, *and elsewhere, and fine porches as at Stanton,* 79, *while the typical East Suffolk mills, like those at Syleham,* 77, *Saxtead Green, see title page, and Pettaugh,* 80, *were on the whole somewhat larger and had higher round houses, often with three floors, and as high as the white painted mill body itself. The oldest is Drinkstone Mill, dating from* 1685, *which stands close to Woolpit Mill,* 81, *and most unusually turns clockwise. The difference in character of post mills which is effected by such things as the shape of the porch and the position of the fantail is well illustrated here. The striking gear for the patent sails can be seen outside the mill to the left of the porch at Syleham Mill, which still grinds corn. See also* 1, 11 *to* 15, 17, 21, 22, *and* 82.

81

82

83, 84

*The difference in character of post mills is effected by such things as the shape of the porch and the position of the fantail. Thorpeness Mill, Suffolk, 82, was erected at Aldringham in 1803. It ground corn until it was moved in 1924 and adapted to pump water for the village of Thorpeness. The roundhouse is modern. The striking gear for the patent sails can be seen outside the mill to the left of the porch. South Essex had a tradition in smock mills of which Upminster Mill, 83, 84, is typical, especially the cap. Built about 1800 it is an unusually large mill for its height and was bought for preservation by the Essex County Council just before the war. The sack loading device hanging down from the stage is unusual. The deep petticoat to the cap is an excellent feature. See also* 30.

85, 86, 87

88, 89, 90

**Yorkshire.** *Only two mills now work by wind in Yorkshire, and of these the five-sailed mill at Seaton Ross, 86, is the last mill in the county to work with roller reefing sails, 42. It stands on the site of an old post mill, some of whose timbers are incorporated in the present structure.*

**Derbyshire.** *Cat and Fiddle Mill, Dale Abbey, 87, is the last mill to work by wind in Derbyshire. It is situate close to the Nottinghamshire border and with its small roundhouse and covered platform at the top of the ladder it is typical of the post mills of both counties. It has a stone bearing for the 'neck' journal of the windshaft, a practice not uncommon in French Flanders to-day. It is owned and kept in working order by the Stanton Ironworks.*

**Nottinghamshire.** *The tower mills of Nottinghamshire are first cousins to those of Lincolnshire with their ogee caps and sails fixed to a cross instead of a poll-end, like that at Coddington, 88. The tower was probably raised when patent sails were fitted, as was often the case when it was no longer necessary to stop the mill to set the sails.*

**Leicestershire.** *The Leicestershire tower mills were also mainly of the Lincolnshire type. This one at Harby, 89, was wrecked in a gale shortly before the war and only one mill in the county is now at work by wind.*

**Lincolnshire.** *There are more corn mills at work in Lincolnshire than any other county at present. The post mill at Friskney Tofts, 93, built in 1730, was blown down in 1939, and only one is now at work by wind; while there are no smock mills at work at all. The last survivor of the early tower mills at Lutton Gowts, 90, was built in 1746 and was dismantled in 1942. It stood on an artificial mound which was probably the site of a still earlier post mill. The tower mill is more highly developed in Lincolnshire than anywhere else and a number of multi-sailed mills were built. Of these six-sailers were most favoured and could, if necessary, be run with four, two or even three sails as at Metheringham, 91. The last mill to be built in Lincolnshire was Trader Mill, Sibsey, 92, in 1877. It was built by Saundersons, of Louth, on the site of an old post mill. See also 20, 23, 28, 45, 47, 49, 50, 54, 58.*

**Norfolk.** *Not all the Norfolk mills have the typical Norfolk boatshaped cap, 40. Denver Mill, 94, has an ogee cap with horizontal instead of vertical boarding, while Diss Mill, 95, had a domed cap of large size with an interesting framing, 55. But Norfolk is best known for its marsh mills on the Broads, 67. See also 2, 3, 6, 5, 52, 76.*

**Cambridgeshire.** *Although the date of erection is unknown Bourn Mill, 96, was standing in 1636, and is the oldest in England. It was bought and repaired by Mr. Alfred Bossom, M.P., with money raised by the exhibition of Epstein's Genesis, and was presented to the Cambridge Preservation Society. Its small size and straight pitch roof mark it as an old mill. Starting life as a smock mill, Downfield Mill, Soham, 97, was wrecked in a storm and rebuilt as an eight-sided tower mill. The cap and fantail arrangement are typical of the local tradition. The last of the large marsh mills, characteristic of the Fens, to work by wind stands at Soham Mere, 98, built by the father of Mr. Hunt, 63, seventy-one years ago; it was hoped to preserve it as a landmark. The scoop wheel, which lifts the water and is housed in the casing behind the mill, is over 28 feet in diameter, and the sails are about 80 feet span. See also 7, 8. 9. 10, 16, 44, 46.*

**Hertfordshire.** *Hormead mills, 99, 100, stand side by side. The post mill had a sixteen-sided wooden roundhouse; the smock mill was built in the 1860's. It was an old joke to say that windmills should not be built close together in case there was not enough wind to run both! No mills now work by wind in Hertfordshire, which does not seem to have had any traditional type of its own.*

91, 92, 93

97, 98, 99

103, 104, 105

**Surrey.** *The oldest working windmill in the country is a post mill standing at Outwood, Surrey, 101, and was built in 1665. It is also the last mill in the county to work by wind. Here it is being fitted with a pair of sails brought from a demolished post mill at Forncett End, Norfolk. The other pair were presented to it by the Society for the Protection of Ancient Buildings.*

**Sussex.** *Although as a rule fantails were not favoured on post mills in Sussex, Argos Hill Mill, 102, has one mounted on the end of the tail pole. Small shutters in the outer half of the leading edges of the sails, and leading boards on the inner half were characteristic Sussex practice. Its built-up post of pitch pine instead of solid oak is unusual. Only one post mill is now at work in the county. Earnley Mill, 103, was the last smock mill to work in Sussex. It has two spring sails and two cloth-spread common sails, quite a usual arrangement at one time. Polegate Mill, 104, was the last of the*

94, 95, 96

100, 101, 102

106, 107, 108

*Sussex tower mills to work by wind. It is unusual in being partly tiled and, like some other Sussex mills, has a five-bladed fantail.*

**Kent.** *The smock mill was predominant in Kent, with a cap like the top of a post mill and Headcorn Mill, 105, can be considered typical with its low wooden stage. The use of this can be seen at Kingsdown Mill, 106, which had two spring and two common sails which were set from the stage; it also had a unique seven-bladed fantail. It was moved from Farningham in 1880 at a cost of £800 and last worked in 1928. Union Mill, Cranbrook, 107, which is 72 feet high to the ridge of the cap, was built in 1814 by Humphrey of Cranbrook, and is still at work, the finest smock mill in England. The wheel on the left side of the cap at one time carried an endless chain for turning the cap by hand. Black Mill, Barham Downs, 108, was built in 1834 by John Holman, of Canterbury, and is a very good example of Holman's tradition in mill building. It stands on the site of an earlier mill. See also 19, 56.*

# Glossary

## The post mill

| | |
|---|---|
| **The Post mill** | *is so called because the body containing the machinery is mounted on a* |
| **post.** | *This is supported by four diagonal* |
| **quarterbars** | *whose lower ends rest on two* |
| **cross trees** | *placed at right angles to each other.* |
| **The crown tree** | *is socketed on to the top of the post on which it turns.* |
| **The side girts** | *rest on either end of the crown tree and on them the framing of the body is built up.* |
| **The sheers** | *run from the breast to the tail of the mill below the bottom floor of the mill and on either side of the post.* |
| **The collar** | *is fixed below the sheers and encloses the post forming a bearing to steady the body of the mill.* |
| **The tailpole** | *extends from the back of the sheers to the ground and is used to turn the body of the mill to face the sails into the wind.* |
| **The round house** | *encloses the substructure of the mill and forms a storage place, but is usually of no structural significance.* |
| **The petticoat** | *is a downward extension of the weather-boarding of the body over the top of the round-house.* |
| **The brakewheel** | *is so called because a contracting brake acts in its rim. It is mounted on* |
| **the windshaft,** | *which carries the sails and it drives* |
| **the wallower,** | *which is the first driven gear in the mill.* |
| **A sunk post mill** | *is one with its substructure buried in the ground.* |

## The tower mill

| | |
|---|---|
| **The tower mill** | *consists of a fixed tower containing the machinery while only the top or* |
| **cap** | *is turned to face the sails into the eye of the wind.* |
| **The dust floor** | *is the top floor of a tower mill excluding dust and dirt from the mill. Below it is* |
| **the bin floor** | *to which the grain is raised before being ground on* |
| **the stone floor** | *between the stones. Afterwards it passes down by spouts to* |
| **the meal floor** | *into sacks or meal bins.* |
| **A smock mill** | *is a tower mill with a polygonal wooden tower having a resemblance to a countryman's smock.* |
| **Truck wheels** | *serve to centre the cap in relation to the tower.* |
| **The curb** | *is the track on the top of the tower on which the cap turns round. It is said to be a dead curb if blocks are attached to the bottom of the cap and a live curb if rollers are used instead of blocks.* |
| **A shot curb** | *consists of rollers running in two concentric rings between tracks on the bottom of the cap and the top of the tower.* |
| **A mill is tail winded** | *if the wind strikes the sails from behind.* |

## The composite mill

| | |
|---|---|
| **A composite mill** | *consists of a postmill body on a short tower.* |

## Winding the mill

| | |
|---|---|
| **The fan tail** | *is an automatic device to keep the sails into the eye of the wind.* |

## Windmill sails

| | |
|---|---|
| **The whip** | *is the backbone of the sail, which is built up of* |
| **bars** | *mortised into the whip and laths connecting the bars.* |
| **The hemlath** | *connects the bars at their outer ends.* |
| **A stock** | *carries two sails, the whips being bolted and strapped to it.* |
| **The weather** | *of a sail can be compared to the twist of a propeller.* |
| **The leading board** | *is fixed to the leading edge of a single-sided sail.* |
| **The common sail** | *is cloth spread and is the earliest type now used in England.* |
| **The spring sail** | *has shutters like a venetian blind instead of sail cloth. These are spring controlled and when open* |
| **spill the wind.** | |

| | |
|---|---|
| **The roller reefing sail** | *has roller blinds instead of shutters. They are connected by webbing straps called* |
| **listings** | *and are simultaneously opened and closed by bars called* |
| **airpoles** | *which are all connected by cranks, levers and a spider coupling to* |
| **the striking rod** | *This passes through the wind shaft and can be moved backwards and forwards from ground level by means of a chain, thus opening and closing the roller blinds.* |
| **The patent sail** | *has a similar remote control, which is self regulating and shutters are used instead of blinds.* |
| **The poll end or canister head** | *is the double socket at the end of the windshaft through which the stocks are passed.* |
| **The cross** | *is an alternative method of fixing sails. Stocks are dispensed with, whips are increased in size and called* |
| **backs** | *and are bolted to the cross, which is keyed on to the windshaft just in front of the* |
| **neck** | *or front journal bearing.* |

## Windmill gearing

| | |
|---|---|
| **A compass arm wheel** | *is one whose spokes are mortised through the shaft on which it is mounted.* |
| **A clasp arm wheel** | *has spokes forming a square at the centre through which the shaft passes.* |
| **A lantern pinion** | *is a gear wheel having top and bottom flanges connected by staves which serve as cogs and it is driven by* |
| **A trundle wheel** | *whose crudely shaped cogs are often little better than pegs; this is the primitive type of gear.* |
| **The tail wheel** | *is mounted at the rear of the windshaft in a post mill and drives a pair of stones.* |
| **The upright shaft** | *is introduced if the drive to the stones is indirect. At the top is the wallower and below is* |
| **the great spur wheel** | *which drives the stones through spur pinions known as* |
| **stone nuts.** | *If the stones are over driven (or over-drift) the nuts are mounted on* |
| **quants** | *which drive* |
| **the runner stones,** | *the grinding taking place between these and the stationary* |
| **bedstones.** | *If the stones are under driven (or under-drift) the nuts are mounted on* |
| **the stone spindles** | *on top of which the runner stones are mounted.* |
| **A cog box** | *is a wooden jig used to cut out the blanks for wooden cogs for gears.* |

## The stones

| | |
|---|---|
| **The horse** | *is the stand supporting the grain hopper and stands on a* |
| **vat** | *or casing surrounding the stones. From it is hung the inclined* |
| **shoe** | *down which the grain falls to the stones. The flow of grain is maintained by* |
| **the damsel** | *which knocks the shoe and the volume of flow is controlled by* |
| **the gate** | *in the shoe. Grain falls into* |
| **the eye** | *of the runner stone, passes into* |
| **the furrows** | *of the stones, and is ground as it passes outwards.* |
| **The bridge tree** | *is the beam carrying the thrust bearing for the stone spindle. It is hinged and by raising and lowering it alters the gap between the stones and is known as* |
| **tentering.** | *This enables the degree of fineness of the meal to be regulated.* |
| **Peak stones** | *are solid stones quarried in the Peak District of Derbyshire.* |
| **French stones** | *are built up of stones quarried in France.* |
| **Blue stones or cullin stones** | *are solid German stones shipped from Cologne.* |

## Machines

| | |
|---|---|
| **The bolter** | *was used for separating flour from bran by beating it through woollen bolting cloth. In a* |
| **wire machine** | *it was brushed through wire gauze.* |
| **The groat machine** | *separates groats from dust and husk.* |

*109. Map showing the location of windmills*

# Index to Windmills mentioned in the book

# Index